A BRIEF INTRODUCTION OF ELECTRONICS (1)

Purpose of the book

The word electronics is mostly used in day to day life. It is the word which commonly used and rarely understood. This book is just a trial to make a common man familiar with electronics. It is written in such a way to avoid scientific definitions and formulae. It is useful to all whether you have a scientific background or not. If you have a desire to know the basic building blocks of electronics and ready to feel the scope and excitement of electronics, it is the right book for you. It is also very- very useful for all students who are going to opt electronics in their college life. It may be a good start for a pre university student.

Preface

This book is written for everyone. It is introduction of electronics. There is neither scientific vocabulary nor definition is used in entire book. There is no any pre requirement of mathematics or science basics. This book reveals the basics, scope and excitement of electronics in our day to day life. For simplicity this complete book is divided into many sub sections. Introduction part deals with the identification of electronics in everyday life. Subsequent sections tell about the building blocks that are essential for an electronic device. The last section reveals the impact of electronics in our day to day life.

Index

Introduction

Electronics is an emerging branch of science. In our day to day life, we come across many machines, devices and appliances which are the gifts of electronics. In our daily life we use T.V, computer, digital note book, remote controller to change T.V channels, digital watches, electronic scooty and traffic signals all of these are the result of electronics development. In our general life we willingly or unwillingly use electronics. Whether it is medical equipment, juice mixer or grinder, fan's speed controller or air conditioner uses electronics principle.

Here a genuine quest appears, how to categories the equipments or devices are electronic in nature or not? In many cases we see that, we are unable to categories or differentiate between electronic and electrical devices. Electronics and electrical is not opposite to each other but complimentary.

We can understand this with a very simple activity. Suppose we have set of battery, a bulb, wire and a switch. If we connect these as follows

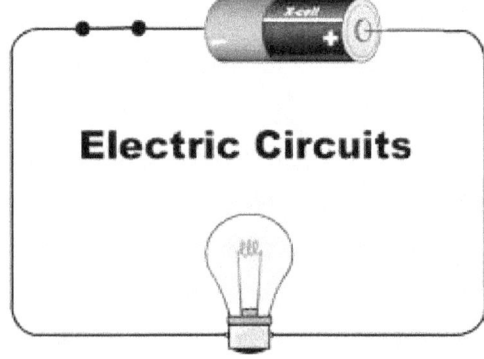

Fig. 1.1 electrical circuit

When switch on the bulb glows. Whenever switch is off the bulb do not glows. The above activity is completely an electrical phenomenon in nature.

But what should be done if we want to control the brightness of the bulb. In this situation electronics comes into picture.

In electrical we have no any control over the flow of current. But in electronics we can control the flow of current. By controlling the current we can control the brightness of bulb.

Suppose if we introduce a knob which can control the current the previous circuit will be electronic circuit. At this stage we are not going to deal with the constituents of the knob. It will form our subject matter of the next lessons. However if you have slightest knowledge of resistor, we can also have a variable resistor as knob.

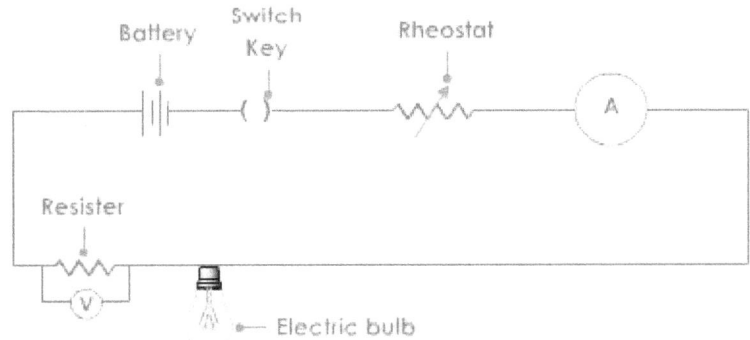

Fig. 1.2 electronic circuit

In general life we use the above mentioned knob. It is used in fans, AC and mixture to control the speed and change the mode of operation.

Building Blocks of Electronics

Before explaining about the building blocks which are used to control the current in a particular circuit, here we will familiarize with some physical quantities which will be generally used in sub sequent topics. Here our aim is not to be master of these terms or have a scientific definition but to have a general idea about them.

Resistor and Resistivity

Resistor is anything which resists the flow of current or consumes the current when current passes through it.

In a general way everything around us is a resistor. Even our human body is a resistor. Some resistors have negligible resistance, some have very high and others in between them.

We can categories them in good resistors or bad resistors. Good resistors are generally termed as bad conductors of electricity or insulators while bad resistors are good conductor of electricity or conductors.

In our first activity if we break the wire and at this place join alluminium wire the bulb glows this shows that alluminium has less resistance or consumption to electricity and is a good conductor. But if we join the ends with a wood stick the bulb do not glows. This proves that the wood is a bad conductor of electricity.

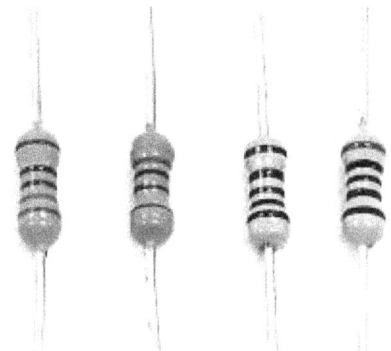

Fig. 1.3 resistors

Unit of resistance is ohm (Ω). It represents quantitatively the resistance of any substance. The measuring instrument of resistance is ohmmeter.

Resistivity

It is the resistance of unit length resistor having unit cross sectional area. Its unit is ohm.m (Ωm) .

$$R = \frac{\rho L}{A}$$

ρ = resistivity
L = length
A = cross sectional area

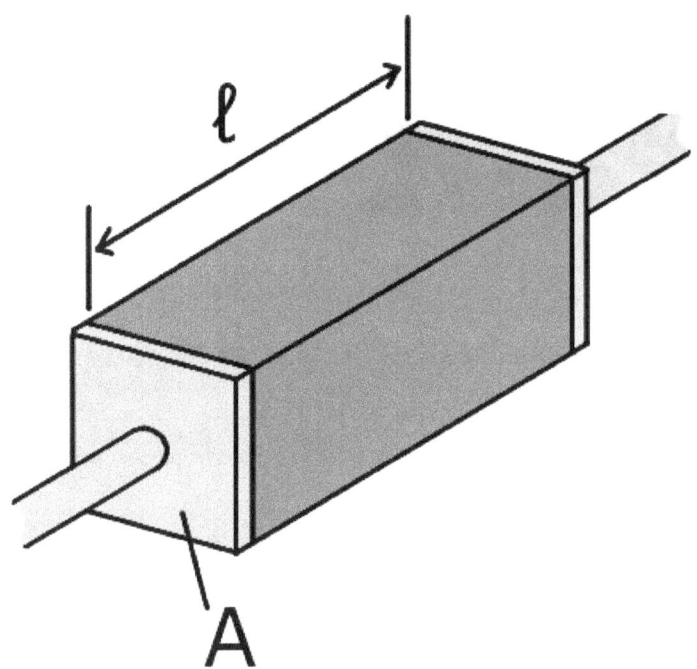

Fig. 1.4 unit length and unit cross sectional resistor

Resistivity of any resistor does not depend on its shape, size but on it is the intrinsic property of a particular material. This can be understood by a simple example, if we have a bulk of 1000kg of copper and a wire of copper having weight 10mg both will have the same resistivity.

Capacitor

Capacitor is an important component in an electronic circuit. When it is introduced in a circuit, it serves as a pot to collect charges or current coming from battery or source up to its capacity and thereafter let the excess current to pass through it. But when supply stops, it starts supplying stored current or charge in opposite direction.

Fig. 1.5 different types of capacitors

Capacitor is also a special type of resistor. It only allows to passes AC currents through it and blocks the DC current. Its unit is farad (F).

Inductor

It is also a special type of resistor. Suppose we have a thin wire of any material like copper. This is a resistor. If wire is wound on a rod in many turns, then this wounded resistor can be called inductor.

Fig. 1.6 symbol of inductor

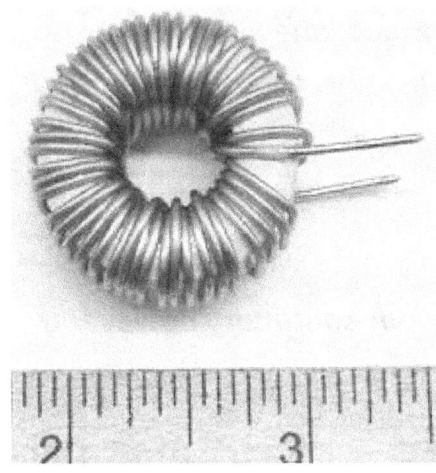

Fig. 1.7 an inductor

It is measured in hennery (h) and represented by "L".

The tailoring material: p-type and n-type semiconductor

So far we have studied that every material has some resistance. It may be a good conductor or insulators. There are some materials whose resistance lies between these two and are generally referred as semiconductors. Si (silicon) and Ge (germanium) are naturally occurring semiconductors while, some are formed by doping or introducing some impurities in a pure element. Pure semiconductors are termed as intrinsic semiconductors while doped are known as extrinsic semiconductor.

Doping provides us to make materials of our desired level of conductivity and we are able to control the conductivity of a material as well as current flowing through it. If we increase conductivity of a particular material, there is less loss of current.

Formation of p-type semiconductor

In modern periodic table there are 18 groups. In these groups all118 elements are arranged whether they are naturally occurring or synthesized in an labs. It is also assumed that the whole universe is composed of these elements only.

p-type semiconductor is formed when 14th group elements are doped with 13th group of elements. As we know that the valency of 14th group is 4 and of thirteen groups are 3. Here it must be clear that valency only refers to the no. of electrons of a particular atom which are free to take part in a chemical reaction.

The following topic can be escaped without affecting our continuity however if you want to have more about valency you can go through it.

HOW TO FIND VALENCY FROM ELECTRONIC CONFIGURATION:

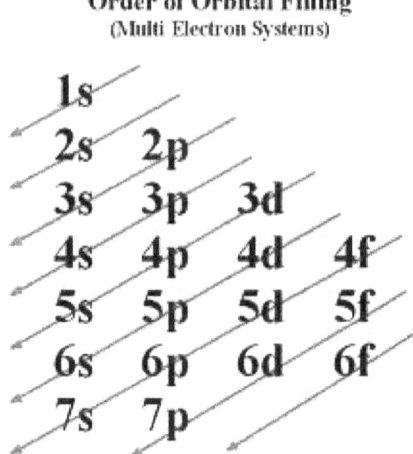

Fig. 1.8 filling of electrons in orbits

We can find the valency of a particular element in simple steps. Before using these steps we must know that the filling of electron is started from 1s then 2s and go on succession with 2p, 3s, 3p, 4s, 3d, 4p, 5s, 4d, 5p, 6s, 4f, 5d, 6p, 7s, 5f, 6d, 7p, 6f and so on as shown in fig. 1. Secondly we must also care that the max no. of electron in a particular shell like s, p, d, and f is limited i.e.

Shell	no. of max electron allowed
S	2
P	6
d	10

Working step:

Example 0.1: find the valency of sodium?

Solution:

Step: 1 find the electronic configuration

      The atomic no. of sodium is 11. So, there are 11 electrons in its neutral atom.

Therefore filling of electrons will be as follows $1s^2$, $2s^2$, $2p^6$, $3s^1$.

Note: here we find that the max no. of electron can be 2 in 3s shells but already 10 electrons have been filled in 1s, 2s and 2p, so only 1 electron is filled in 3s.

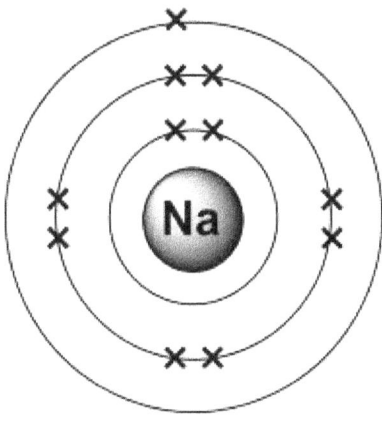

Fig. 1.9 electronic configuration of Na

Step: 2   in the outer most shell there is only one electron. Hence its valency is one.

Note: valence electron refers to the no. of electrons residing in the outer most shell and only these electrons take part in a particular reaction. If there are more than 4 electron in the outer most orbit. For finding valency we must subtract the no of electrons from 8 (as it is necessary to complete octet). But this time valency is in negative due to receiving electron from other atom electron has negative charge.

We can see it in the next example:

Example 0.2: find the valency of oxygen?

Solution:

Step: 1 find the electronic configuration

The atomic no. of oxygen is 8. So, there are 8 electrons in its neutral atom.

Therefore filling of electrons will be as follows $1s^2$, $2s^2$, $2p^4$.

Note: Here we find that the max no. of electron can be 6 in 2p shell but already 4 electrons have been filled in 1s and 2s so only 4 electrons is filled in 2p.

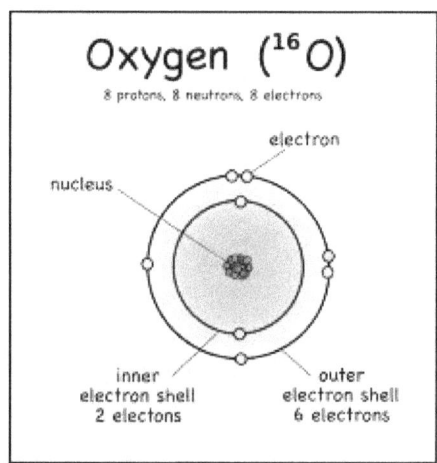

Fig. 1.10 electronic configuration of Oxygen

Step: 2 in the outer most shell there are 6 electrons (2 in 2s and 4 in 2p as outer most orbit is 2) only , which is greater than 4. Hence its valency 6-8=-2.

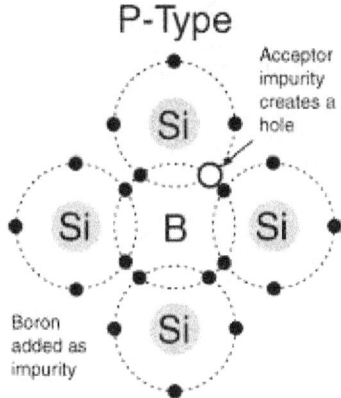

Fig. 1.11 formation of p-type semiconductor

When an element of valency 3 is doped in the lattice of 14th group element in the lattice a hole or void place is created due to less no. of valency and the electrons of other bonded 14th group element move to fill the void and this result in conduction. It is due to an empty space generally termed as hole.

Formation of n-type semiconductor

In the formation of n-type semiconductor the 14th group element is doped with a 15th group element whose valency is 5 i.e. one extra electron than the 14 th group. This extra electron does not take part in bonding and randomly moves in the entire crystal and is responsible for conductivity

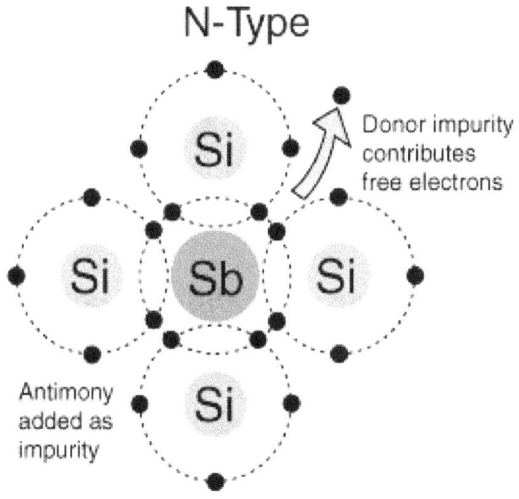

Fig. 1.12 formation of n-type semiconductor

Now we have an idea of conductivity in p-type and n-type semiconductors. as per our requirements of conductivity we can increase or decrease the no. of conduction carriers i.e. hole in p-type and electrons in n-type. That's why these are termed as **tailoring materials**.

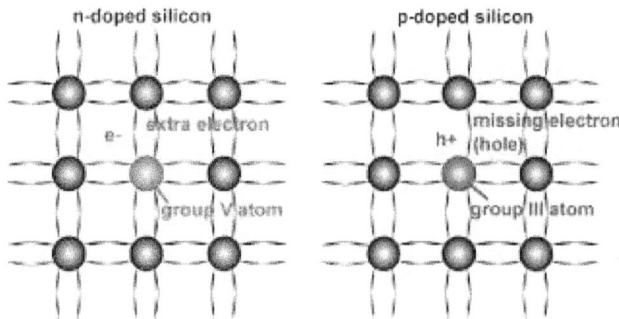

Fig. 1.13 p and n-type semiconductor

Diode: the p-n junction

Before moving to the diode the fundamental and basic building blocks of electronics like brick to a building, we will concentrate on the following two topics which are prominent for the formation of diode:-

a. **drift**

b. **diffusion**

Drift

In literary meaning drift refers to the dragging something from somewhere to other place. Here it has the same meaning the slight change is that the dragging is done by applying voltage or potential difference.

It is the motion of charged particle like electron due to the potential difference.

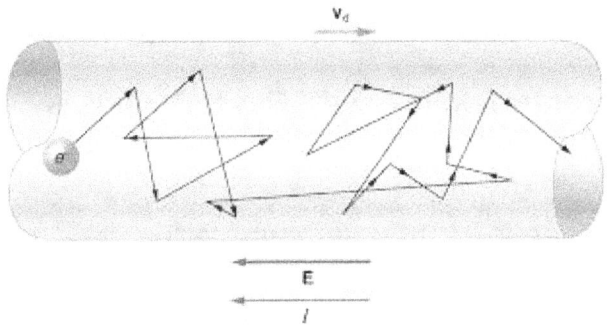

Fig. 1.14 drift of electron in a conductor

Diffusion

In drift the motion is due to potential gradient but in diffusion it is due to concentration gradient. We know that the rate refers to the variation of anything with respect to the time similarly gradient refers to the variation of anything with respect to the position.

Suppose we have a container. In this container there is a separating permeable boundary and half part is completely filled while other half is empty.

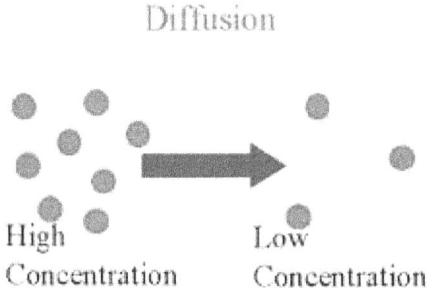

Fig. 1.15 diffusion process

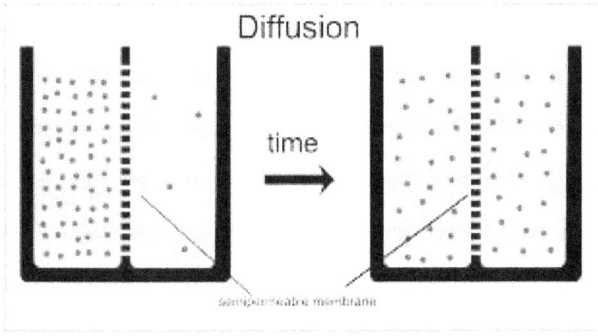

Fig. 1.16 diffusion

Then the matter will flow from higher concentration to lower. This technique is used in RO (reverse osmosis) water purifier.

A very simple example of diffusion can be seen in a class room where the diffusion of dust particle from duster to the table takes place. After some days we are unable to completely clean the upper surface due to diffused particle of chalk in the table.

Formation of p-n junction

In the formation of p-n junction diffusion has a great role. After the formation of p-n junction, when it is connected into the circuit the second term drift comes into picture.

When p-type and n-type materials are atomically connected p-n junction is formed. Here it is most importantly noted that the connection must be at atomic level. We can't form p-n junction just by joining the p-type material with n-type material.

Now we will see actually what happens when p-type and n-type materials come into contact at atomic level.

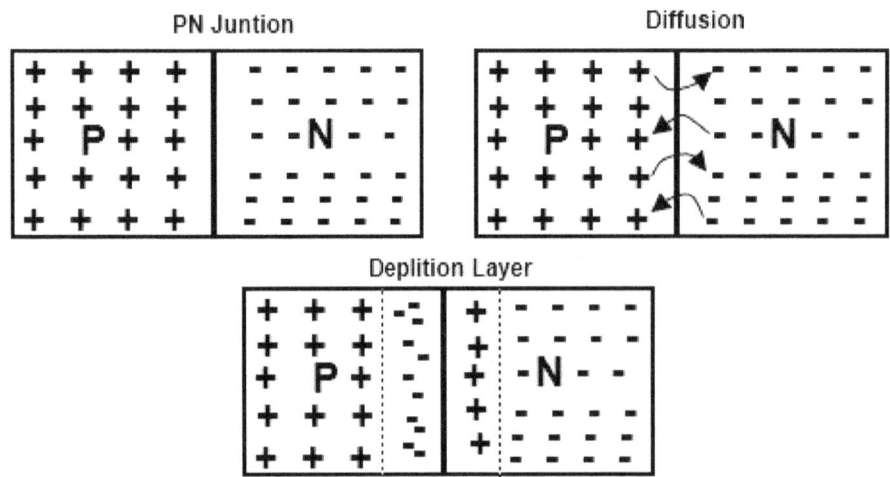

Fig. 1.17 diffusion in p-n junction

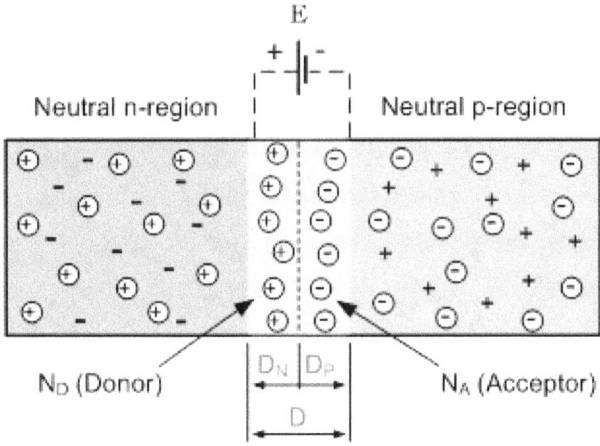

Fig. 1.18 depletion region

We see that in p-type material holes are in majority and are termed a majority carrier while in n-type material electrons are in majority carrier.

When p-type and n-type are combined atomically, a concentration gradient is created. Due to this holes start flowing from p section to n and electrons from n section to p. both of these carriers combine at the junction and form an immobile region generally termed as depletion region. As this barrier formed *holes start accumulating at the boundary of depletion region towards the n side and electrons towards the p side, as they have less energy to cross this region.*

Biasing of diodes

When diodes or p-n junction is connected to a power supply, there are two ways in which the diode can be connected. When p terminal is connected to the positive terminal of the diode, it is in forward biased. When p terminal is connected to the negative terminal of battery and n section to the positive of batter, it is reverse biased.

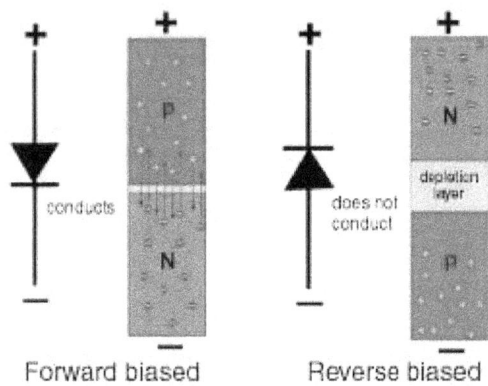

Forward biased   Reverse biased

Fig. 1.19 biasing of diodes

Forward biasing

When forward biasing is applied drift phenomenon comes into picture. Due to drift the hole accumulated towards n section tends to move towards p section and electrons from p to n section. This reduces depletion region. But after some time this region becomes thin having only immobile charges through which current can easily pass through.

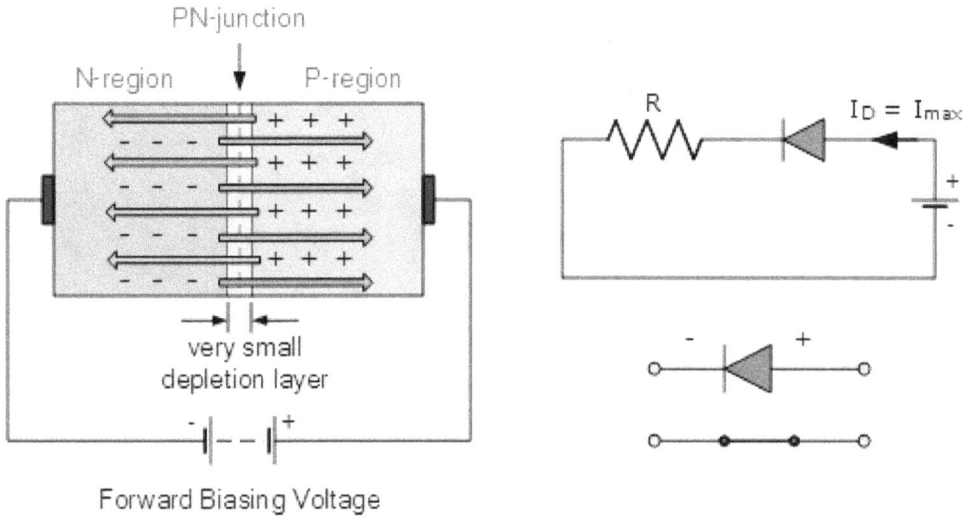

Fig. 1.20 forward biasing of diode

Reverse biasing

As in forward biasing, in reverse biasing drift also comes into picture. But in this biasing electrons in p section combine with holes of p section and repel the electrons coming from negative terminal as p section is connected with negative terminal. Similarly, in n section the holes at the boundary of potential barrier combine with free electrons in n section and repel holes coming from positive terminal. Due to this the depletion region or generally termed as potential barrier is widen and makes current almost impossible to pass through this.

Therefore in reverse biasing diodes do not conduct. It serves as an open circuit i.e. a broken circuit.

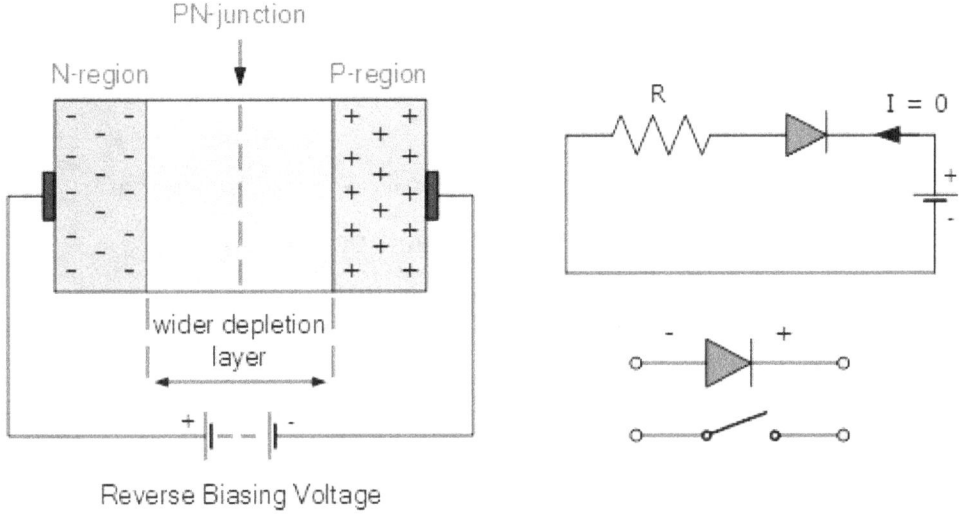

Fig. 1.21 reverse biasing of diode

Voltage consumption of diodes

When diode is used as a component in a circuit and connected in forward biased it consumes some voltage as this voltage is required to overcome the depletion region. This consumption depends on the material we used during the formation of p-type and n-type material.

It is 0.7 volt for silicon diode and 0.3 for germanium diode.

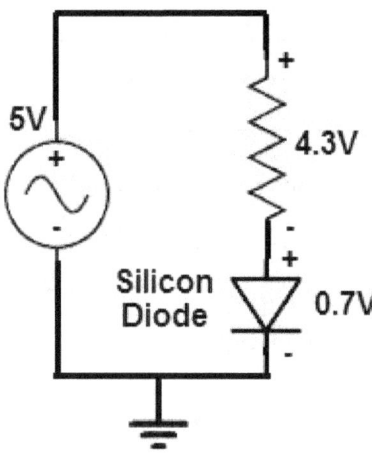

Fig. 1.22 voltage consumption of silicon diode

In the above circuit it is seen that the electric bulb could get only 4.3 volt and 0.7 volt is consumed by silicon diode.

After the knowledge about the behavior of a general purpose diode, we can use it in infinite no. of operations like in clipper circuit and clamper circuit etc.

LED

LED used in general life are actually light emitting diodes. So, they must be used in forward biased as to produce light. For an LED if its negative terminal (generally the shorter leg) is connected to the positive terminal of battery or voltage source will never glow the LED.

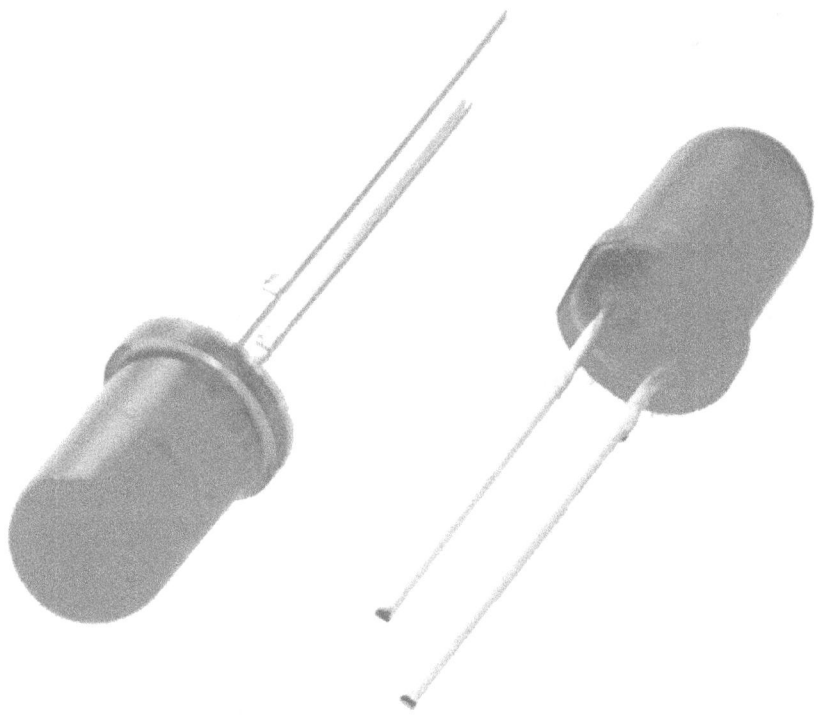

Fig. 1.23 LED

Transistors

The diode is a two terminal device. What configuration may be if we try to form a three terminal device using p-type and n-type semiconductors?

There are two possible configurations first is to sandwich p-type material between two n-type and second to sandwich n-type between two p-type materials. This can be seen as follows:

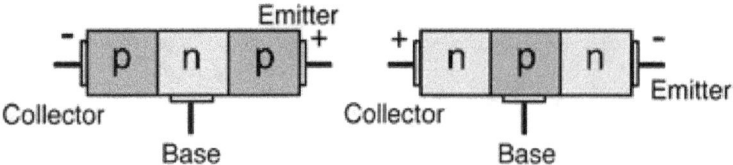

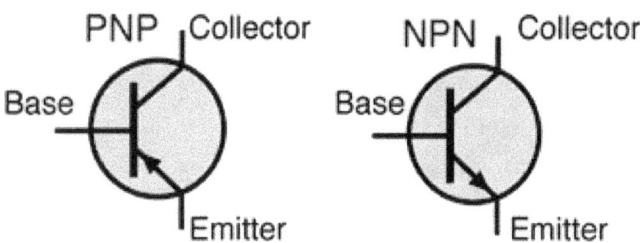

Fig. 1.24 symbol and schematic diagram of transistor

These two configurations are used to form pnp and npn transistors. However there is slight change in the doping profile of material of three sections and their width.

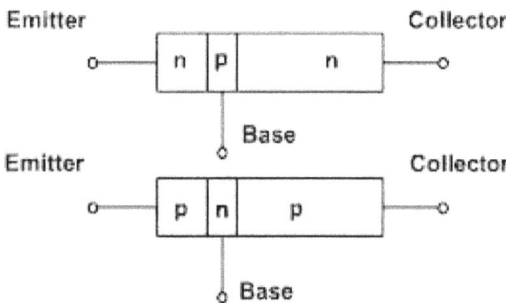

Fig. 1.25 transistors

In this configuration the current enters at emitter section, passes through emitter-base junction, base-collector junction and finally reaches into collector from where we can get it as an output.

As current enters at emitter its resistance is kept least and size or width in between the base and collector. Its doping profile is highly doped to keep its conductivity high. Base area is kept very thin almost 1/150 times the emitter and least doped as to easily cross this section by current. The collector has larger area as it has to collect the current and resistance is kept high.

Fig. 1.26 different types of transistors

Why the name transistor?

Transistor is a current controlled device. In the transistor we see that there is transfer of resistance from very low i.e. emitter to very high of collector. Suppose the resistance of emitter is approx. 1^{-3} ohm and of collector is 1 ohm. When 1 ampere current passes through the emitter its input voltage will be as per ohm's law

$V = IR = 10^{-3}$ volt.

When this current enters into the collector section the output voltage becomes

$V = IR = 1$ volt.

Here we can see that the output voltage is 1000 times the input voltage. We can also control this voltage gain by just changing the doping profile of base.

We can understand the function of emitter, base and collector in a simple way.

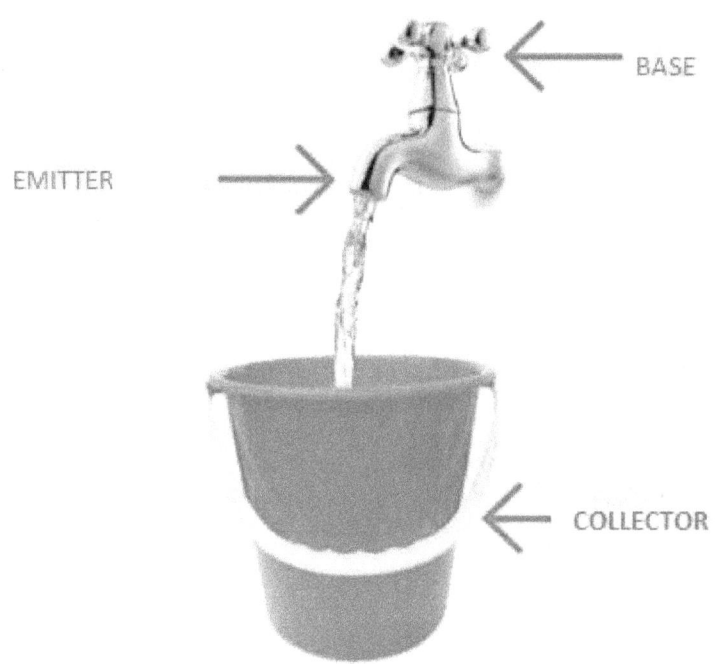

Fig. 1.27 showing concept of base, emitter and collector

Suppose we have a tap of water supply, the supplying pipe can be treated as emitter, the con trolling knob as a base and the collecting bucket as collector. As a transistor is a current controlled device, we can control the current to get desired level of voltage for our use.

Can we get a transistor by joining two diodes back to back?

During the formation of diodes we have seen that the p-type and n-type materials must be joined at atomic level. If we join two diodes back to back doping profile will vary as they can't be automatically connected so, it is not possible to get a desired transistor by just joining two diodes back to back.

Modes and configuration of operation of transistors

Transistor can have three configurations namely common base, common emitter and common collector.

In common base the base is common to both emitter and collector.

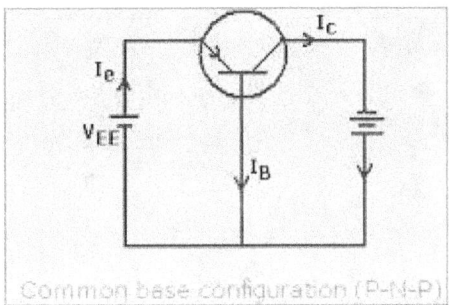

Fig.1.28

In common collector the collector terminal is common to both emitter and base.

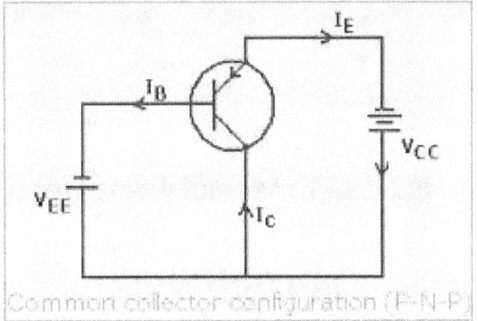

Fig. 1.29

In common emitter the emitter terminal is kept commom.

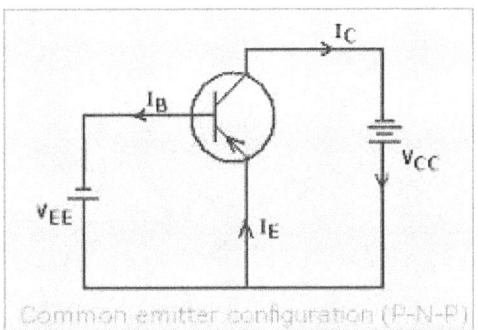

Fig. 1.29

For clarity we will go through only common base configuration. Now this configuration can be used in four different modes which are

a. cut off mode

b. saturation mode

c. forward active mode

d. reverce active mode

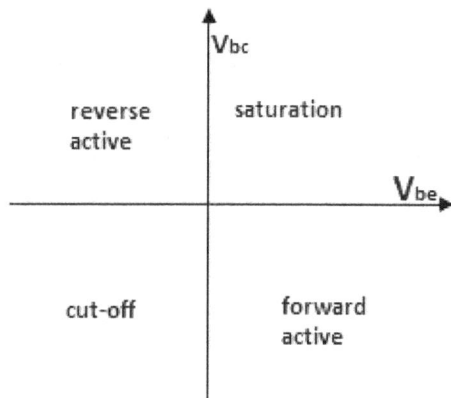

Fig. 1.30 modes of transistor

Active mode

When base emitter junction is forward biased and emitter collector junction is reverse biased the transistor is in active mode.

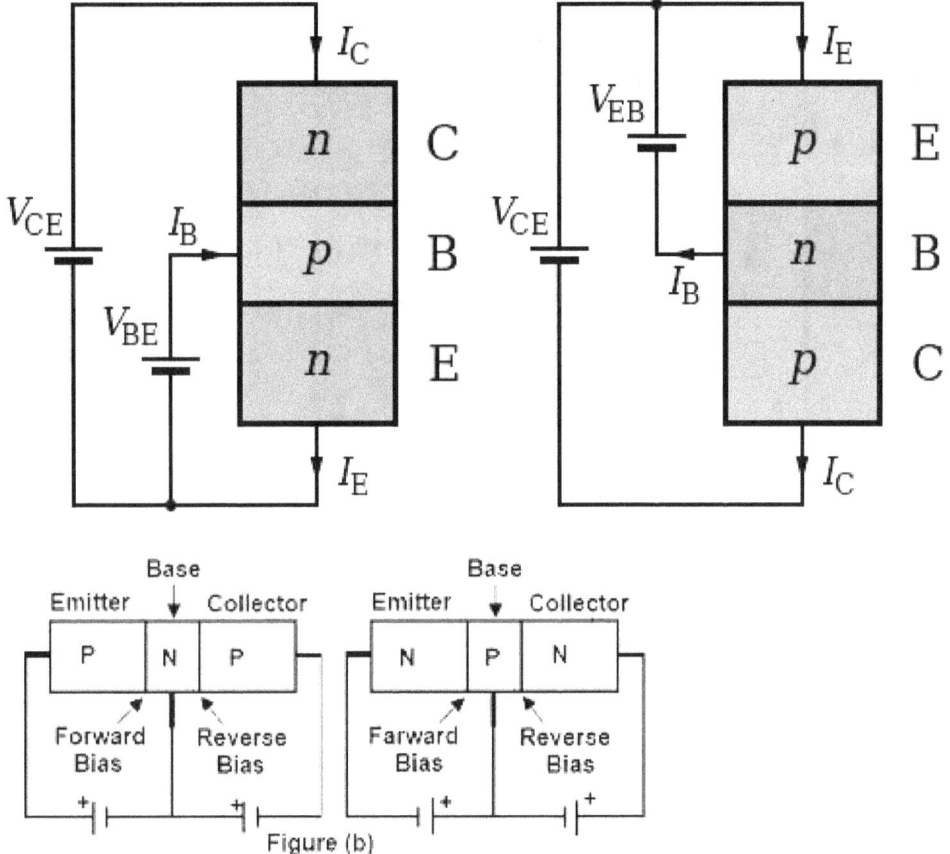

Fig. 1.31 active mode in common base configuration

Cut-off mode

When base emitter junction and collector emitter junction both are reverse biased the transistor is in cut off mode.

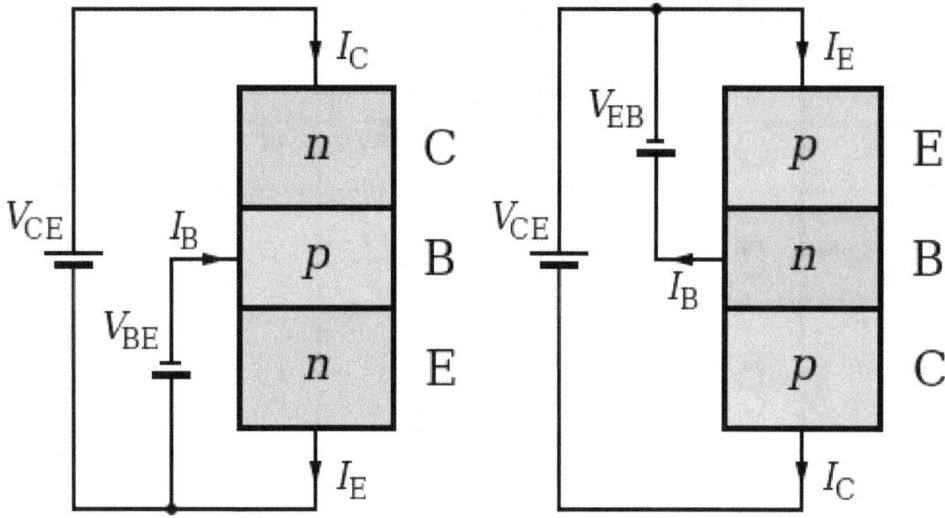

Fig. 1.32 cut-off mode in cb

Saturation mode

When both junctions are in forward biased transistor is in saturation mode.

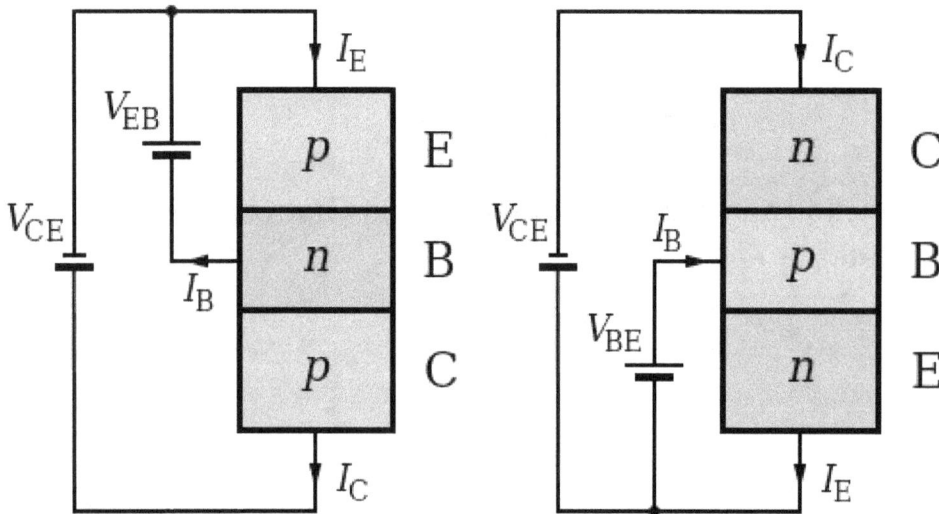

Fig. saturation mode in cb

Each and every mode has its own application during designing a particular device. Here we are interested in only giving a brief introduction to the building blocks. We are not going inside the uses and application of these modes.

There have been many modifications in transistors. FET is field effect transistor. It is a voltage controlled device. In FET emitter is termed as source, base as gate and collector as drain. However the working principle solely depends on drift and diffusion.

There have been many more devices designed with slightest modification such as MOSFET i.e. metal oxide field effect transistor. UJT is uni junction transistor and controlled by single carrier. SCR and different types of diodes like varactor diode, zener diode are developed as a two terminal devices. Each has its unique uses and requirement. Zener diode is generally used in reverse biased and can be used as a voltage regulator.

In all electronic devices the basic concept of electrons, holes, drift and diffusion remains same.

How transistors and diodes are running the world?

In everyday life most widely and commonly used chips like SIM card of our mobile phones are nothing but a huge combination of diodes and transistors.

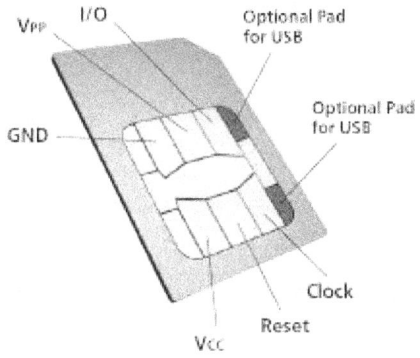

Memory chips and integrated circuits which constitute our mother board and hard disc of our computer and electronic devices are the result of joining different resistors, diodes and transistors.

 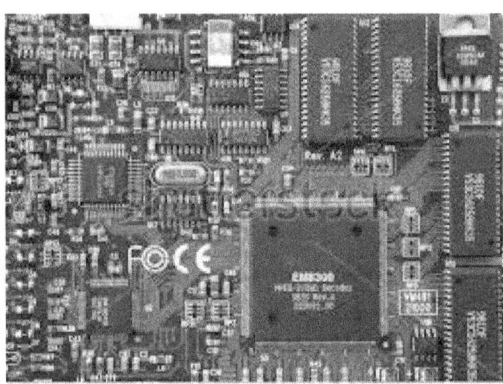

These chips may be DTL, TTL, RTL etc i.e. diode-transistor logic, transistor-transistor logic and resistor-transistor logic respectively.

The speed of a chip, memory card or microprocessor solely depends on the purification and doping level of p-type and n-type materials.

NOTES